H. Smeathman

Some Account of the Termites, Which are Found in Africa and Other Hot Climates

In a Letter from Mr. Henry Smeathman, of Clement's Inn, to Sir Joseph

Banks, Bart. P. R. S.

H. Smeathman

Some Account of the Termites, Which are Found in Africa and Other Hot Climates
In a Letter from Mr. Henry Smeathman, of Clement's Inn, to Sir Joseph Banks, Bart.
P. R. S.

ISBN/EAN: 9783337309138

Printed in Europe, USA, Canada, Australia, Japan

Cover: Foto ©berggeist007 / pixelio.de

More available books at **www.hansebooks.com**

XI. *Some Account of the* Termites, *which are found in* Africa *and other hot Climates. In a Letter from Mr.* Henry Smeathman, *of* Clement's Inn, *to Sir* Jofeph Banks, *Bart. P. R. S.*

Read February 15, 1781.

SIR,
Clement's Inn,
Jan. 23, 1781.

OF a great many curious parts of the creation 1 met with on my travels in that almoft unknown diftrict of Africa called Guinea, the TERMITES, which by moft travellers have been called WHITE ANTS, feemed to me on many accounts moft worthy of that exact and minute attention which I have beftowed upon them.

The amazingly great and fudden mifchief they frequently do to the property of people in tropical climates, makes them well known and greatly feared by the inhabitants.

The fize and figure of their buildings have attracted the notice of many travellers, and yet the world has not hitherto been furnifhed with a tolerable defcription of them, though their contrivance and execution fcarce fall fhort of human ingenuity and prudence ; but when we come to confider the wonderful oeconomy of thefe infects, with the good order of their fubterraneous cities, they will appear foremoft on the lift of the wonders of the creation, as moft clofely imitating mankind in provident induftry and regular government.

T 2
You

You had barely time to fee and to admire fome of their buildings in New Holland, and have been pleafed to fay, you think an accurate account of them would meet a favourable reception from the Royal Society. That which I now have the honour to prefent to you, is accurate and faithful as far as it goes. I have kept as clofe to my fubject as was in my power, without being obfcure, or falling fhort of my intention; and though I have given only the heads of what I could draw from my memorandums on the fubject, they will probably be found fufficiently defcriptive and hiftorical for the bounds of a letter.

The fagacity of thefe little infects is fo infinitely beyond that of any other animals I have ever heard of, that it is poffible the accounts I have here communicated would not appear credible to many, without fuch vouchers and fuch corroborating teftimony as I am fortunately able to produce, and are now before you. There are alfo many living witneffes in England to moft of the extraordinary relations that I have given, fo that I hope to have full credit for fuch remarks as no one but myfelf has probably had time and opportunities enough to make, and which are not fufceptible of demonftration, except in thofe places where the infects are found.

Such as they are, I beg leave to lay them, with all diffidence and humility, before you and that illuftrious Body of which you are Prefident; and if they fhould in a fmall degree meet with approbation, I fhall be exceedingly fatisfied.

Thefe infects are known by various names. They belong to the TERMES of LINNÆUS, and other fyftematical naturalifts.

By the Englifh, { In the windward parts of Africa they are called *Bugga Bugs.* In the Weft Indies, *Wood Lice, Wood Ants,* or *White Ants.*

By

By the French,
{
~~At Senegal, *Vague-Vagues.*~~
~~In the Weſt Indies, *Poux de Bois,* or *Four-*~~
~~*mis Blanches.*~~
}

By the French, At Senegal, *Vague-Vagues.* In the Weſt Indies, *Poux de Bois,* or *Fourmis Blanches.*

By the Bolms, or Sherbro people, in Africa, *Scantz.*

By the Portugueze in the Brazils, *Coupée* or *Cutters,* from their cutting things in pieces.

By this latter name and that of *Piercers* or *Eaters,* and ſimilar terms, they are diſtinguiſhed in various parts of the tropical regions.

The following are the ſpecific differences, given by Dr. SOLANDER, of ſuch inſects of this *genus* as I have obſerved and collected.

1. TERMES *bellicoſus* corpore fuſco, alis fuſceſcentibus : coſtâ ferrugineâ, ſtemmatibus ſubſuperis oculo propinquis, puncto centrali prominulo.

2. TERMES *mordax* nigricans, antennis pedibuſque teſtaceis, alis fuliginoſis : areâ marginali dilatatâ : coſtâ nigricante, ſtemmatibus inferis oculo approximatis, puncto centrali impreſſo.

3. TERMES *atrox* nigricans, ſegmentis abdominalibus margine pallidis, antennis pedibuſque teſtaceis, alis fuliginoſis : coſtâ nigrâ, ſtemmatibus inferis, puncto centrali impreſſo.

4. TERMES *deſtructor* nigricans, abdominis lineâ laterali luteâ, antennis teſtaceis, alis hyalinis : coſtâ luteſcente, ſtemmatibus ſubſuperis, puncto centrali obliterato.

5. TERMES *arborum* corpore teſtaceo, alis fuſceſcentibus : coſtâ luteſcente, capite nigricante, ſtemmatibus inferis oculo approximatis, puncto centrali impreſſo.

The Termites are repreſented by LINNÆUS as the greateſt plagues of both Indies, and are indeed every way between the

2.   Tropics

Tropics fo deemed, from the vaft damages they caufe, and the loffes which are experienced in confequence of their eating and perforating wooden buildings, utenfils, and furniture, with all kinds of houfehold-ftuff and merchandize, which are totally deftroyed by them, if not timely prevented; for nothing lefs hard than metal or ftone can efcape their moft deftructive jaws.

They have been taken notice of by various travellers in different parts of the torrid zone; and indeed where numerous, as is the cafe in all equinoctial countries and iflands that are not fully cultivated, if a perfon has not been incited by curiofity to obferve them, he muft have been very fortunate who, after a fhort refidence, has not been compelled to it for the fafety of his property.

Thefe infects have generally obtained the name of Ants, it may be prefumed, from the fimilarity in their manner of living, which is, in large communities that erect very extraordinary nefts, for the moft part on the furface of the ground, from whence their excurfions are made through fubterraneous paffages or covered galleries, which they build whenever neceffity obliges, or plunder induces, them to march above ground, and at a great diftance from their habitations carry on a bufinefs of depredation and deftruction, fcarce credible but to thofe who have feen it. But notwithftanding they live in communities, and are like the ants omnivorous; though like them at a certain period they are furnifhed with four wings, and emigrate or colonize at the fame feafon; they are by no means the fame kind of infects, nor does their form correfpond with that of *Ants* in any one ftate of their exiftence, which, like moft other infects, is changed feveral times.

The Termites refemble the Ants alfo in their provident and diligent labour, but furpafs them as well as the Bees, Wafps,

3   Beavers,

Beavers, and all other animals which I have ever heard of, in the arts of building, as much as the Europeans excel the leaft cultivated favages. It is more than probable they excel them as much in fagacity and the arts of government ; it is certain they fhew more fubftantial inftances of their ingenuity and induftry than any other animals ; and do in fact lay up vaft maga-zines of provifions and other ftores ; a degree of prudence which has of late years been denied, perhaps without reafon, to the Ants [1].

Such however are the extraordinary circumftances attending their oeconomy and fagacity, that it is difficult to determine, whether they are more worthy of the attention of the curious and intelligent part of mankind on thefe accounts, or from the ruinous confequences of their depredations, which have defervedly procured them the name of *Fatalis* or *Deſtruʃtor.*

As this is the cafe, it is a little furprifing that an accurate natural hiftory of thefe wonderful infects has not been attempted long fince ; efpecially as, according to bosman (who wrote the beginning of this century) in his defcription of the Coaft of Guinea, fome curious cicumftances relative to them muft have been known. According to that gentleman, the *King* was fuppofed to be as large as a Cray-fifh [2]. This, though a bad comparifon, is pretty near the truth in refpect to the fize of the female, who is the *Common Mother* of the community ;

[1] Though Ants have no occafion to lay up ftores for winter in cold climates, they certainly muft and do carry great quantities of provifions into their nefts to feed the young brood ; and moft probably provide fome before hand for fear of accidents, which might be fatal to the young ones, who, like all infects in the caterpillar ftate, are very voracious, and cannot bear difappointments of long duration.

[2] bosman's Guinea, p. 260.

and, according to the mode we have adopted from time immemorial in fpeaking of Ants and Bees, the QUEEN.

Thefe communities confift of one *male* and one *female* (who are generally the *common parents* of the whole, or greater part, of the reft), and of three orders of infects, apparently of very different fpecies, but really the fame, which together compofe great commonwealths, or rather monarchies, if I may be allowed the term.

The great LINNÆUS, having feen or heard of but two of thefe orders, has claffed the genus erroneoufly ; for he has placed it among the *Aptera*, or infects without wings ; whereas the chief order, that is to fay, the infect in its perfect ftate, having four wings without any fting, it belongs to the *Neuroptera* ; in which clafs it will conftitute a new genus of many fpecies [3].

The different fpecies of this genus refemble each other in form, in their manner of living, and in their good and bad qualities : but differ as much as birds in the manner of building their habitations or nefts, and in the choice of the materials of which they compofe them.

There are fome fpecies which build upon the furface of the ground, or part above and part beneath, and one or two fpecies, perhaps more, that build on the ftems or branches of trees, fometimes aloft at a vaft height.

[3] I have no doubt, from the account and figures given of the European Termes Pulfatorius, or Death Watch, by the illuftrious BARON DE GEER, in his feventh volume of *Memoires pour fervir à l' Hiftoire des Infectes*, that in their perfect ftate they have wings, and fwarm or emigrate, and live in a manner analogous to thofe of hot climates ; for they feem to have quite the external form of the exotic Termes, that is to fay, of the firft and third order. DE GEER, Memoires, tom. VII. p. 45. pl. IV. fig. 1, 2, 3, & 4.

Of every species there are three orders; firft, the working infects, which, for brevity, I fhall generally call *labourers*; next the fighting ones, or *foldiers*, which do no kind of labour; and, laft of all, the winged ones, or *perfect infects*, which are male and female, and capable of propagation. Thefe might very appofitely be called the *nobility* or *gentry*, for they neither labour, or toil, or fight, being quite incapable of either, and almoft of felf-defence. Thefe only are capable of being elected kings or queens; and nature has fo ordered it, that they emigrate within a few weeks after they are elevated to this ftate, and either eftablifh new kingdoms, or perifh within a day or two.

The *Termes bellicofus* being the largeft fpecies is moft re-markable and beft known on the Coaft of Africa. It erects immenfe buildings of well-tempered clay or earth, which are contrived and finifhed with fuch art and ingenuity, that we are at a lofs to fay, whether they are moft to be admired on that account, or for their enormous magnitude and folidity. It is from the two lower orders of this, or a fimilar fpecies, that LINNÆUS feems to have taken his defcription of the *Termes Fatalis*; and moft of the accounts brought home from Africa or Afia of the white Ants are alfo taken from a fpecies that are fo much alike in external habit and fize, and build fo much in their manner, that one may almoft venture to pronounce them mere variations of the fame fpecies.

The reafon that the larger Termites have been moft remarked is obvious; they not only build larger and more curious nefts, but are alfo more numerous, and do infinitely more mifchief to mankind. When thefe infects attack fuch things as we would not wifh to have injured, we muft confider them as moft perni-cious; but when they are employed in deftroying decayed trees and fubftances which only incumber the furface of the earth, they may be juftly fuppofed very ufeful, and for the reafon tha they are in one fenfe moft pernicious, they are in the other moft ufeful. In this refpect they refemble very much the common Flies, which are regarded by mankind in general as noxious, and at beft as ufelefs beings in the creation; but this is certainly for want of confideration. There are not probably in all nature animals of more importance, and it would not be difficult to prove, that we fhould feel the want of one or two fpecies of large quadrupeds, much lefs than of one or two fpecies of thefe defpicable-looking infects. Mankind in general are fenfible that nothing is more difagreeable, or more pefti-ferous, than putrid fubftances; and it is apparent to all who have made obfervation, that thofe little infects contri-bute more to the quick diffolution and difperfion of pu-trefcent matter than any other. They are fo neceffary in all hot climates, that even in the open fields a dead animal or fmall putrid fubftance cannot be laid upon the ground two mi-nutes before it will be covered with Flies and their Maggots, which inftantly entering quickly devour one part, and per-forating the reft in various directions, expofe the whole to be much fooner diffipated by the elements. Thus it is with the Ter-mites; the rapid vegetation in hot climates, of which no idea can be formed by any thing to be feen in this, is equalled by as great a degree of deftruction from natural as well as accidental caufes[4]. It feems apparent, that when any thing whatever is arrived at its laft degree of perfection, the Creator has decreed it fhall

---

[4] The Guinea grafs, which is fo well known and fo much efteemed by our planters in the Weft Indies, grows in Africa thirteen feet high upon an average, which height it attains in about five or fix months; and the growth of many other plants is as quick.

be

be totally deſtroyed as ſoon as poſſible, that the face of nature may be ſpeedily adorned with freſh productions in the bloom of ſpring or the pride of ſummer: ſo when trees, and even woods, are in part deſtroyed by tornadoes or fire, it is wonderful to obſerve, how many agents are employed in haſtening the total diſſolution of the reſt (5); but in the hot climates there are none ſo expert, or who do their buſineſs ſo expeditiouſly and effectually, as theſe inſects, who in a few weeks deſtroy and carry away the bodies of large trees, without leaving a particle behind, thus clearing the place for other vegetables, which ſoon fill up every vacancy; and in places, where two or three years before there has been a populous town, if the inhabitants, as is frequently the caſe, have choſen to abandon it, there ſhall be a very thick wood, and not the veſtige of a poſt to be ſeen, unleſs the wood has been of a ſpecies which, from its hardneſs, is called *iron wood.*

My general account of the Termites is taken from obſervations made on the *Termes bellicoſus,* to which I was induced by the greater facility and certainty with which they could be made.

The neſts of this ſpecies are ſo numerous all over the iſland of Bananas, and the adjacent continent of Africa, that it is ſcarce poſſible to ſtand upon any open place, ſuch as a rice plantation, or other clear ſpot, where one of theſe buildings is not be ſeen within fifty paces, and frequently two or three are to be ſeen almoſt cloſe to each other. In ſome parts near Senegal, as mentioned by Monſ. ADANSON, their number, magnitude, and cloſeneſs of ſituation, make them appear like the villages of the natives (6): and you have yourſelf ſeen them perhaps ſtill more numerous, though not ſo large, in New Holland. Theſe

(5) See STILLINGFLEET's Tracts.

(6) " But of all the extraordinary things I obſerved, nothing ſtruck me more " than certain eminences, which, by their height and regularity, made me take

" them

These buildings are usually termed hills, by natives as well as strangers, from their outward appearance, which is that of little hills more or less conical, generally pretty much in the form of sugar loaves, and about ten or twelve feet in perpendicular height above the common surface of the ground.[7] [8] [9], tab. VII. fig. 1.

These

" them at a distance for an assemblage of negroes huts or a considerable village,
" and yet they were only the nests of certain insects. They are round pyramids
" from eight to ten feet high, upon nearly the same base, with a smooth surface
" of rich clay, excessively hard and well built." adanson's Voyage to Senegal,
8vo, p. 153—337. Voyage de Senegal, 4to, p. 83 and 99.

Note, What Mr. adanson says of the opening which gives ingress and regress
is manifestly a mistake, arising from the natural conclusion that those insects had
some way out and in to their nests, without examining where it was. It will
appear by this account, that they have many thousand ways out and in, but all
subterraneous.

(7) jobson, in his History of Gambia, says, " The Ant hills are remarkable
" cast up in those parts by Pismires, some of them twenty foot in height, of
" compasse to contayne a dozen men, with the heat of the sun baked into that
" hardnesse, that we used to hide ourselves in the ragged toppes of them, when
" we took up stands to shoot at deere or wild beasts." purchas's Pilgrims, vol.
II. p. 1570.

(8) " The Ants make nests of the earth about *twice the height of a man.*"
bosman's Description of Guinea, p. 276—493.

(9) The labourers are not quite a quarter of an inch in length; however, for
the sake of avoiding fractions, and of comparing them and their buildings with
those of mankind more easily, I estimate their length or height so much, and
the human standard of length or height, also to avoid fractions, at six feet; which
is likewise above the height of men. If then one labourer is $=$ to one-fourth of
an inch $=$ to six feet, four labourers are $=$ to one inch in height $=$ 24 feet,
which multiplied by 12 inches, gives the comparative height of a foot of their
building $=$ 288 feet of the building of men, which multiplied by 10 feet, the supposed average height of one of their nests is $=$ 2880 of our feet, which is 240 feet
more than half a mile, or near five times the height of the great pyramid; and, as it

Thefe hills continue quite bare until they are fix or eight feet high; but in time the dead barren clay, of which they are compofed, becomes fertilized by the genial power of the elements in thefe prolific climates, and the addition of vegetable falts and other matters brought by the wind; and in the fecond or third year, the hillock, if not over-fhaded by trees, becomes, like the reft of the earth, almoft covered with grafs and other plants; and in the dry feafon, when the herbage is burnt up by the rays of the fun, it is not much unlike a very large hay-cock.[10]

Every one of thefe buildings confifts of two diftinct parts, the exterior and the interior.

The exterior is one large fhell in the manner of a dome, large and ftrong enough to inclofe and fhelter the interior from the viciffitudes of the weather, and the inhabitants from the attacks of natural or accidental enemies. It is always, therefore, much ftronger than the interior building, which is the habitable part divided with a wonderful kind of regularity and contrivance into an amazing number of apartments for the refidence of the *king* and *queen,* and the nurfing of their nu-

is proportionably wide at the bafe, a great many times its folid contents. If to this comparifon we join that of the time in which the different buildings are erected, and confider the Termites as raifing theirs in the courfe of three or four years, the immenfity of their works fets the boafted magnitude of the antient wonders of the world in a moft diminutive point of view, and gives a fpecimen of induftry and enterprize as much beyond the pride and ambition of men as St. Paul's Cathedral exceeds an Indian hut.

(10) See a figure of one of thofe nefts in SALMON's Univerfal Traveller, in the map of Gambia, where it is called a Pifmire Hill: there is alfo a figure of one of the labouring infects; but as the hill is reprefented below all proportion, and the infect rather larger than life, it gives no idea of the building. I have not been able to find out from what author SALMON took this figure; and it is the only one I have met with.

merous.

merous progeny ; or for magazines, which are always found well filled with ftores and provifions.

I ſhall forbear at this time entering into a very minute account of the infide of thefe wonderful buildings, as the bare recital might appear tedious; though I flatter myfelf, that when I have an opportunity of communicating it to the publick at large, the readers will follow me through an exact defcription of them with pleafure.

Thefe hills make their firft appearance above ground by a little turret or two in the ſhape of fugar loaves, which are run a foot high or more [11]. Soon after, at fome little diftance, while the former are increafing in height and fize, they raife others, and fo go on increafing the number and widening them at the bafe, till their works below are covered with thefe turrets, which they always raife the higheft and largeft in the middle, and by filling up the intervals, between each turret, collect them as it were into one dome.

They are not very curious or exact about thefe turrets, except in making them very folid and ftrong, and when by the junction of them the dome is compleated, for which purpofe the turrets anfwer as fcaffolds, they take away the middle ones entirely, except the tops (which joined together make the crown of the cupola) and apply the clay to the building of the works within, or to erecting frefh turrets for the purpofe of raifing the hillock ftill higher; fo that no doubt fome part of the clay is ufed feveral times, like the boards and pofts of a mafon's fcaffold.

[11] Some of thefe turrets are reprefented in the view of their hills, (tab. VII. fig. 3.). I have feen turrets on the fides of thefe nefts four or five feet high (tab. VII. fig. 1. a. a. a.).

3   When

When thefe hills are at about little more than half their height, it is always the practice of the wild bulls to ftand as centinels upon them, while the reft of the herd is ruminating below (tab. VII.). They are fufficiently ftrong for that purpofe, and at their full height anfwer excellently as places to look out. I have been with four men on the top of one of thefe hillocks. Whenever word was brought us of a veffel in fight, we immediately ran to fome Bugga Bug hill, as they are called, and clambered up to get a good view, for upon the common furface it was feldom poffible to fee over the grafs or plants, which, in fpite of monthly brufhings, generally prevented all horizontal views at any diftance.

The outward fhell or dome is not only of ufe to protect and fupport the interior buildings from external violence and the heavy rains; but to collect and preferve a regular degree of genial warmth and moifture which feems very neceffary for hatching the eggs and cherifhing the young ones.

The *royal chamber*, which I call fo on account of its being adapted for, and occupied by, the *king* and *queen*, appears to be in the opinion of this little people of the moft confequence, being always fituated as near the center of the interior building as poffible, and generally about the height of the common furface of the ground, at a pace or two from the hillock. It is always nearly in the fhape of half an egg or an obtufe oval within, and may be fuppofed to reprefent a long oven (tab. VIII. fig. 1. and 2.).

In the infant ftate of the colony, it is not above an inch or thereabout in length; but in time will be increafed to fix or eight inches or more in the clear, being always in proportion to the fize of the *queen*, who, increafing in bulk as in age, at length requires a chamber of fuch dimenfions.

This

This fingular part would bear a long defcription, which I fhall not trouble you with at prefent, and only obferve, that its floor is perfectly horizontal; and in large hillocks, fometimes an inch thick and upward of folid clay. The roof alfo, which is one folid and well-turned oval arch, is generally of about the fame folidity, but in fome places it is not a quarter of an inch thick, this is on the fides where it joins the floor (tab. VIII. fig. 1. a. a.), and where the doors or entrances are made level therewith at pretty equal diftances from each other (tab. VIII. fig. 2. and 4. b. b.)

Thefe entrances will not admit any animal larger than the foldiers or labourers, fo that the *king*, and the *queen* (who is, at full fize, a thoufand times the weight of a *king*) can never poffibly go out.

The royal chamber, if in a large hillock, is furrounded by an innumerable quantity of others of different fizes, fhapes, and dimenfions ; but all of them arched in one way or another, fometimes circular, and fometimes elliptical or oval.

Thefe either open into each other or communicate by paffages as wide, and being always empty are evidently made for the foldiers and attendants, of whom it will foon appear great numbers are neceffary, and of courfe always in waiting.

Thefe apartments are joined by the magazines and nurferies. The former are chambers of clay, and are always well filled with provifions, which to the naked eye feem to confift of the rafpings of wood and plants which the Termites deftroy, but are found in the microfcope to be principally the gums or infpiffated juices of plants. Thefe are thrown together in little maffes, fome of which are finer than others, and refemble the fugar about preferved fruits, others are like tears of gum, one

quite

quite tranfparent, another like amber, a third brown, and a fourth quite opaque, as we fee often in parcels of ordinary gums.

Thefe magazines are intermixed with the nurferies, which are buildings totally different from the reft of the apartments : for thefe are compofed entirely of wooden materials, feemingly joined together with gums. I call them the nurferies becaufe they are invariably occupied by the eggs, and young ones, which appear at firft in the fhape of labourers, but white as fnow. Thefe buildings are exceeding compact, and divided into many very fmall irregular-fhaped chambers, not one of which is to be found of half an inch in width (tab. VIII. fig. 5.). They are placed all round the royal apartments, and as near as poffible to them.

When the neft is in the infant ftate, the nurferies are clofe to the royal chamber ; but as in procefs of time the queen enlarges, it is neceffary to enlarge the chamber for her accommodation ; and as fhe then lays a greater number of eggs, and requires a greater number of attendants, fo it is neceffary to enlarge and encreafe the number of the adjacent apartments ; for which purpofe the fmall nurferies which are firft built are taken to pieces, rebuilt a little farther off a fize bigger, and the number of them encreafed at the fame time.

Thus they continually enlarge their apartments, pull down, repair, or rebuild, according to their wants, with a degree of fagacity, regularity, and forefight, not even imitated by any other kind of animals or infects that I have yet heard of.

There is one remarkable circumftance attending the nurferies, which I muft not at this time omit. They are always found flightly overgrown with *mould* (tab. VIII. fig. 6), and plentifully fprinkled with fmall white globules about the fize of a fmall pin's head. Thefe at firft I took to be the eggs ; but,

on bringing them to the microfcope, they evidently appeared to
be a fpecies of mufhroom, in fhape like our eatable mufhroom in the young ftate in which it is pickled (tab. VIII. fig. 7.).
They appear, when whole, white like fnow a little thawed
and then frozen again, and when bruifed feem compofed of an
infinite number of pellucid particles, approaching to oval forms
and difficult to feparate ; the mouldinefs feems likewife to be
the fame kind of fubftance [12].

The nurferies are inclofed in chambers of clay, like thofe
which contain the provifions, but much larger. In the early
ftate of the neft they are not bigger than an hazel-nut, but in
great hills are often as large as a child's head of a year old.

The difpofition of the interior parts of thefe hills is pretty
much alike, except when fome infurmountable obftacle prevents;
for inftance, when the *king* and *queen* have been firft lodged near
the foot of a rock or of a tree, they are certainly built out of the
ufual form, otherwife pretty nearly according to the following
plan.

The royal chamber is fituated at about a level with the furface of the ground, at an equal diftance from all the fides of
the building, and directly under the apex of the hill (tab. VII.
fig. 2. A. A.).

[12] Mr KONIG, who has examined thefe kind of nefts in the Eaft Indies, in an
Effay upon the Termites, read before the Society of Naturalifts of Berlin, con-
jectures, that thefe mufhrooms are the food of the young infects. This fuppofition
implies, that the old ones have a method of providing for and promoting their
growth ; a circumftance which, however ftrange to thofe unacquainted with the
fagacity of thefe Infects, I will venture to fay, from many other extraordinary facts
I have feen of them, is not very improbable.

N. B. Mr. KONIG has not difcovered the magazines of provifions in the nefts
which he opened, as far as I am informed ; but I muft obferve here, that what I
have learned of this gentleman's account was from an extempore tranflation of the
heads of it.

It is on all fides, both above and below, furrounded by what I fhould call the *royal apartments*, which have only labourers and foldiers in them, and can be intended for no other purpofe than for thefe to wait in, either to guard or ferve their common FATHER and MOTHER, on whofe fafety depends the happinefs, and, according to the negroes, even the exiftence of the whole community.

Thefe apartments compofe an intricate labyrinth, which extends a foot or more in diameter from the *royal chamber* on every fide. Here the nurferies and magazines of provifions begin, and, being feparated by fmall empty chambers and galleries, which go round them or communicate from one to the other, are continued on all fides to the outward fhell, and reach up within it two-thirds or three-fourths of its height, leaving an open area in the middle under the dome, which very much refembles the nave of an old cathedral : this is furrounded by three or four very large Gothic-fhaped arches, which are fometimes two or three feet high next the front of the area, but diminifh very rapidly as they recede from thence like the arches of aifles in perfpectives, and are foon loft among the innumerable chambers and nurferies behind them.

All thefe chambers, and the paffages leading to and from them, being arched, they help to fupport one another; and while the interior large arches prevent them falling into the center, and keep the area open, the exterior building fupports them on the outfide.

There are, comparatively fpeaking, few openings into the great area, and they for the moft part feem intended only to admit that genial warmth into the nurferies which the dome collects.

X 2 The

The interior building or affemblage of nurferies, chambers, &c. has a flattifh top or roof without any perforation, which would keep the apartments below dry, in cafe through accident the dome fhould receive any injury and let in water; and it is never exactly flat and uniform, becaufe they are always adding to it by building more chambers and nurferies : fo that the divifions or columns between the future arched apartments re-femble the pinnacles upon the fronts of fome old buildings, and demand particular notice as affording one proof that for the moft part the infects project their arches, and do not make them, as I imagined for a long time, by excavation (tab. VII. fig. 2. B.).

The area has alfo a flattifh floor, which lays over the royal chamber, but fometimes a good height above it, having nurferies and magazines between (tab. VII. fig. 2. c.). It is likewife water-proof, and contrived, as far as I could guefs, to let the water off, if it fhould get in, and run over by fome fhort way into the fubterraneous paffages which run under the loweft apartments in the hill in various directions, and are of an aftonifhing fize, being wider than the bore of a great cannon. I have a memorandum of one I meafured, perfectly cylindrical, and thirteen inches in diameter (tab. VII. fig. 2. D. D.).

Thefe fubterraneous paffages or galleries are lined very thick with the fame kind of clay of which the hill is compofed, and afcend the infide of the outward fhell in a fpiral manner, and winding round the whole building up to the top interfect each other at different heights, opening either immediately into the dome in various places, and into the interior building, the new turrets, &c. or communicating thereto by other galleries of different bores or diameters, either circular or oval.

From every part of thefe large galleries are various fmall pipes or galleries leading to different parts of the building.

Under

Under ground there are a great many which lead downward by floping defcents three and four feet perpendicular among the gravel, from whence the labouring Termites cull the finer parts, which, being worked up in their mouths to the confiftence of mortar, becomes that folid clay or ftone of which their hills and all their buildings, except their nurferies, are compofed.

Other galleries again afcend and lead out horizontally on every fide, and are carried under ground near to the furface a vaft diftance : for if you deftroy all the nefts within one hundred yards of your houfe, the inhabitants of thofe which are left unmolefted farther off will neverthelefs carry on their fubterraneous galleries, and invade the goods and merchandizes contained in it by fap and mine, and do great mifchief, if you are not very circumfpect.

But to return to the cities from whence thefe extraordinary expeditions and operations originate, it feems there is a degree of neceffity for the galleries under the hills being thus large, being the great thoroughfares for all the labourers and foldiers going forth or returning upon any bufinefs whatever, whether fetching clay, wood, water, or provifions ; and they are certainly well calculated for the purpofes to which they are applied, by the fpiral flope which is given them ; for if they were perpendicular the labourers would not be able to carry on their building with fo much facility, as they afcend a perpendicular with great difficulty, and the foldiers can fcarce do it at all. It is on this account that fometimes a road like a ledge is made on the perpendicular fide of any part of the building within their hill, which is flat on the upper furface, and half an inch wide, and afcends gradually like a ftair-cafe, or like thofe roads which are cut on the fides of hills and mountains, that would otherwife

wife be inacceffible : by which, and fimilar contrivances, they travel with great facility to every interior part.

This too is probably the caufe of their building a kind of bridge of one vaft arch, which anfwers the purpofe of a flight of ftairs from the floor of the area to fome opening on the fide of one of the columns which fupport the great arches, which muft fhorten the diftance exceedingly to thofe labourers who have the eggs to carry from the royal chamber to fome of the upper nurferies, which in fome hills would be four or five feet in the ftraighteft line, and much more if carried through all the winding paffages which lead through the inner chambers and apartments.

I have a memorandum of one of thefe bridges, half an inch broad, a quarter of an inch thick, and ten inches long, making the fide of an elliptic arch of proportionable fize ; fo that it is wonderful it did not fall over or break by its own weight before they got it joined to the fide of the column above. It was ftrengthened by a fmall arch at the bottom, and had a hollow or groove all the length of the upper furface, either made purpofely for the inhabitants to travel over with more fafety, or elfe, which is not improbable, worn fo by frequent treading (tab. VII. fig. 2. E. E.).

Thus I have defcribed, as briefly as the fubject would admit, and I truft without exaggeration, thofe wonderful buildings whofe fize and external form have often been mentioned by travellers, but whofe interior and more curious parts are fo little known, that I may venture to confider my account of them as new, which is the only merit it has : for they are conftructed upon fo different a plan from any thing elfe upon the earth, and fo complicated, that I cannot find words equal to the tafk, and muft therefore refer to the different figures, which, however extraordinary, fcarce do juftice to the fubjects.

The nefts before defcribed are fo remarkable on account of their fize, that travellers have feldom, where they were to be feen, taken notice of any other; and have generally, when fpeaking of white Ants, defcribed them as inhabitants of thofe hills. Thofe, however, which are built by the fmaller fpecies of thofe infects, are very numerous, and fome of them exceedingly worth our attention; one fort in particular, which from their form I have named turret nefts. Thefe are a great deal lefs than the foregoing, and indeed much lefs in proportion to the fize of the builders; but their external form is more curious, and their folidity confidered they are prodigious buildings for fo fmall an animal [13].

Thefe buildings are upright cylinders compofed of a well-tempered black earth or clay, about three quarters of a yard high, and covered with a roof of the fame material in the fhape of a cone, whofe bafe extends over and hangs down three or four inches wider than the perpendicular fides of the cylinder, fo that moft of them refemble in fhape the body of a round wind-mill; but fome of the roofs have fo little elevation in the middle, that they are pretty much in the fhape of the top of a full-grown mufhroom (tab. IX. fig. 1.)

After one of thefe turrets is finifhed, it is not altered or enlarged; but when no longer capable of containing the community, the foundation of another is laid within a few inches of it. Sometimes, though but rarely, the fecond is begun before the firft is finifhed, and a third before they have completed the

[13] If their height is eftimated and computed by the fize of the builders, and compared with ours upon the like fcale; each of them is four or five times the height of the monument, and a great many times its folid contents.

fecond

fecond : thus they will run up five or fix of thefe turrets at the foot of a tree in the thick woods, and make a moft fingular group of buildings (tab. IX.).

The turrets are fo ftrongly built, that in cafe of violence they will much fooner overfet from the foundation, and tear up the gravel and folid earth, than break in the middle; and in that cafe the infects will frequently begin another turret and build it, as it were, through that which is fallen ; for they will connect the cylinder below with the ground, and run up a new turret from its upper fide, fo that it will feem to reft upon the horizontal cylinder only (tab. IX. fig. 5.).

I have not obferved any thing elfe about thefe nefts that is remarkable, except the quality of the black brown clay, which is as dark coloured as rich vegetable mould, but burns to an exceeding fine and clear red brick. Within, the whole building is pretty equally divided into innumerable cells of irregular fhapes ; fometimes they are quadrangular or cubic, and fometimes pentagonal; but often the angles are fo ill defined, that each half of a cell will be fhaped like the infide of that fhell which is called the Sea-ear.

Each cell has two or more entrances, and as there are no pipes or galleries, no variety of apartments, no well-turned arches, wooden nurferies, &c. &c. they do not by any means excite our admiration fo much as the hill nefts, which are indeed collections of wonders.

There are two fizes of thefe turret nefts, built by two different fpecies of Termites. The larger fpecies, the *Termes atrox*, in its perfect ftate meafures one inch and three-tenths from the extremities of the wings on the one fide to the extremities on the other (tab. X. fig. 14.). The leffer fpecies, *Termes mordax*, measures meafures only eight-tenths of an inch from tip to tip (tab. X. fig. 10.

The next kind of nefts, built by another fpecies of this genus, the *Termes arborum*, have very little refemblance to the former in fhape or fubftance. Thefe are generally fpherical or oval, and built in trees [14]. Sometimes they are feated between the arms and the ftems of trees, and very frequently may be feen furrounding the branch of a tree at the height of feventy or eighty feet; and (though but rarely of fo large a fize) as big as a very great fugar cafk [15] [16].

They are compofed of fmall particles of wood and the various gums and juices of trees, combined with, perhaps, thofe of the animals, and worked by thofe little induftrious creatures into a pafte, and fo moulded into innumerable little cells of very different and irregular forms, which afford no amufing variety and nothing curious, but the immenfe quantity of inhabitants, young and old, with which they are at all times crouded; on which account they are fought for in order to feed young fowls, and efpecially for the rearing of Turkies. Thefe nefts are very compact, and fo ftrongly attached to the boughs on which they are fixed, that there is no detaching them but by cutting them in pieces, or fawing off the branch; and they will fuftain the force of a tornado as long as the tree on which they are fixed.

[14] The colour of thefe nefts, like that of the roofed turrets, is black, from which, and their irregular furface and orbicular fhape, they have been called *Negro Heads* by our firft writers on the Carribbee Iflands, and by the French, *Tetes des Negres.* See HUNTER's EVELYN's SILVA, p. 17.

I have never been able to difcover what author Mr. EVELYN alludes to in this mention of the Negro Heads.

[15] LONG's Jamaica, vol. III. p. 887.

[16] SLOANE's Jamaica, vol. II. p. 221. and fequel.

This fpecies has the external habit, fize, and almoft the colour, of the *Termes atrox* (tab. X. fig. 21.).

There are fome nefts built in thofe fandy plains which we call, after the Spaniards, *Savannas,* that refemble the hill nefts firft defcribed. They are compofed of a black mud, which is brought from a few inches below the white fand, and are built in the form of an imperfect cone, or bell-fhaped, having their tops rounded. Thefe nefts are generally about four or five feet high [17]. As I faw thefe only in paffing through various Savannahs upon other purfuits, I can fay very little of their interior parts. They feemed to be inhabited by nearly as large infects, differing very little except in colour, which is lighter than that of the *Termites bellicofi.*

Having given fome idea of the nefts, I fhall beg your patient reading of a more particular account of the infects themfelves, which will be exceeding neceffary to a tolerable acquaintance with their oeconomy and management, their manner of building, fighting, and marching, and to a more particular account of their ufes in the creation, and of the vaft mifchief they caufe to mankind.

---

[17] " The nefts of Ants are about four feet wide at the bafe, and two high, " of an hemifpherical form. Though made in loofe fand, they are fo hard as " not to be broken without great efforts; and a laden cart could not break " through.— In October and November they add a new ftory.— The Cochons " de Terre (the Left Ant-eater of Mr. PENNANT) make holes in thefe nefts eight " inches in diameter and fix deep; *and having deftroyed the inhabitants, the neft* " *is abandoned; but fometimes the Ants repair it.*" This laft paragraph feems rather founded on conjecture. Voyage au Cap. par M. L'Abbé DE LA CAILLE, p. 305—356.

OVIEDO alfo fays Ants make hillocks as high as a man.

Among

Among thefe you will find, I muſt confeſs, ſome very
extraordinary relations, and many that do not admit a poſſi-
bility of demonſtration ; ſuch is the deſcription of the form of
an army of the *Termites viarum* marching, and the account
of the regularity uſed by the *Termites bellicoſi* in repairing a
breach in their hills. But the very ſingular facts, of which
you have the proofs before you, are ſufficient I ſhould conceive
to procure me belief for the others. Should any perſon doubt,
I would wiſh them to conſider, that a ſtudent of nature
and nature's laws, in any matter relating thereto, has no temp-
tation to tranſgreſs the bounds of truth. I am very ſenſible,
that the works of the creation, and the order thereof, are
eſtabliſhed in the higheſt wiſdom ; that it is as abſurd to attempt
to exaggerate as to detract from them ; and can only ſerve to
expoſe the ignorance of him who attempts it. Beſides, what
I have here advanced muſt be confirmed or contradicted in two
or three years, ſince it will doubtleſs be examined into by all
the curious who viſit tropical regions.

I have obſerved before, that there are of every ſpecies of
Termites three orders ; of theſe orders the working inſects or
labourers are always the moſt numerous; in the *Termes bellicoſus*
there ſeems to be at the leaſt one hundred labourers to one of the
fighting inſects or ſoldiers. They are in this ſtate about one-
fourth of an inch long, and twenty-five of them weigh about a
grain ; ſo that they are not ſo large as ſome of our ants (tab. X.
fig. 6.). From their external habit and fondneſs for wood, they
have been very expreſſively called *Wood Lice* by ſome people, and
the whole genus has been known by that name, particularly among the French. They reſemble them, it is true,
very much at a diſtance, but they run as faſt or faſter than any

other infects of their fize, and are inceffantly buftling about
their affairs [18].

The fecond order, or foldiers, have a very different form from
the labourers, and have been by fome authors fuppofed to be
the males, and the former neuters; but they are, in fact, the
fame infects as the foregoing, only they have undergone a
change of form, and approached one degree nearer to the perfect ftate. They are now much larger, being half an inch
long, and equal in bulk to fifteen of the labourers (tab. X.
fig. 8).

There is now likewife a moft remarkable circumftance in the
form of the head and mouth; for in the former ftate the mouth is
evidently calculated for gnawing and holding bodies; but in this
ftate, the jaws being fhaped juft like two very fharp awls a
little jagged (tab. X. fig 9.), they are incapable of any thing
but piercing or wounding, for which purpofes they are very
effectual, being as hard as a crab's claw, and placed in a
ftrong horny head, which is of a nut-brown colour, and
larger than all the reft of the body together, which feems to
labour under great difficulty in carrying it: on which account
perhaps the animal is incapable of climbing up perpendicular
furfaces.

The third order, or the infect in its perfect ftate, varies its
form ftill more than ever. The head, thorax, and abdomen,
differ almoft entirely from the fame parts in the labourers and
foldiers; and, befides this, the animal is now furnifhed with four
fine large brownifh, tranfparent, wings, with which it is at the
time of emigration to wing its way in fearch of a new fettle-

(18) ROCHFORT, in the Hiftory of the Carribee Iflands, calls them Wood Lice,
and mentions the deftruction they make, &c. p. 149.

ment [19].

ment (19). In fhort, it differs fo much from its form and appearance in the other two ftates, that it has never been fuppofed to
be the fame animal, but by thofe who have feen it in the
fame neft; and fome of thefe have diftrufted the evidence of
their fenfes. It was fo long before I met with them in the nefts
myfelf, that I doubted the information which was given me by
the natives, that they belonged to the fame family (tab. X. fig.
1.) Indeed we may open twenty nefts without finding one
winged one, for thofe are to be found only juft before the com-
mencement of the rainy feafon, when they undergo the laft
change, which is preparative to their colonization. Add to
this, they fometimes abandon an outward part of their
building, the community being diminifhed by fome accident to me unknown. Sometimes too different fpecies of
the real Ant (Formica) poffefs themfelves by force of a
lodgement, and fo are frequently diflodged from the fame
neft, and taken for the fame kind of infects. This I know
is often the cafe with the nefts of the fmaller fpecies,
which are frequently totally abandoned by the Termites,
and completely inhabited by different fpecies of Ants, Cock-
roaches, Scolopendræ, Scorpions, and other vermin, fond of
obfcure retreats, that occupy different parts of their roomy
buildings; which clearly accounts for your having met with
the real Ants in thofe nefts in New Holland.

---

(19) " There is a fort that frequently flies, having red wings. — This flying
" fort flings up the largeft hills, and is wonderfully nimble and induftrious."
KOLBEIN's Cape of Good Hope, 8vo, vol. II. p. 173.

DAPPER calls the Wood Ants *Acolalan,* and fays it becomes as big as one's
thumb, and then takes wing. Defcription de l'Afrique, folio, p. 459.

In

In the winged ftate they have alfo much altered their fize as well as form. Their bodies now meafure between fix and feven tenths of an inch in length, and their wings above two inches and a half from tip to tip, and they are equal in bulk to about thirty labourers, or two foldiers. They are now alfo fur-nifhed with two large eyes placed on each fide of the head, and very confpicuous; if they have any before, they are not eafily to be diftinguifhed. Probably in the two firft ftates, their eyes, if they have any, may be fmall like thofe of moles; for as they live like thefe animals always under-ground, they have as little occafion for thefe organs, and it is not to be wondered at that we do not difcover them; but the cafe is much altered when they arrive at the winged ftate in which they are to roam, though but for a few hours, through the wide air, and explore new and diftant regions. In this form the animal comes abroad during or foon after the firft tornado, which at the latter end of the dry feafon proclaims the approach of the enfuing rains, and feldom waits for a fecond or third fhower, if the firft, as is generally the cafe, happens in the night, and brings much wet after it [20].

The quantities that are to be found the next morning all over the furface of the earth, but particularly on the waters, is aftonifhing; for their wings are only calculated to carry them

[20] " At night I vifited Mr. HARRISON on board the floop; during the time " we had a dreadful tornado, in which a fort of large flies with long wings came " on board in fuch prodigious numbers, that flying into the flames of the " candles, the table was foon covered with thofe that burnt their wings; and " others, which were not burnt, as they walked along the table fhed their wings, " and then were nothing but fo many perfect large maggots." June 10, 1732. MOOR's Travels, p. 118,

a few

a few hours, and after the rifing of the fun not one in a thou-fand is to be found with four wings, unlefs the morning con-tinues rainy, when here and there a folitary being is feen wing-ing its way from one place to another, as if folicitous only to avoid its numerous enemies, particularly various fpecies of Ants which are hunting on every fpray, on every leaf, and in every poffible place, for this unhappy race; of which probably not a pair in many millions get into a place of fafety, fulfil the firft law of nature, and lay the foundation of a new com-munity.

Not only all kinds of ants, birds, and carnivorous reptiles, as well as infects, are upon the hunt for them, but the inha-bitants of many countries, and particularly of that part of Africa where I was, eat them.[21] [22] [23] [24] [25].

On

(21) Mr. KONIG, in an Effay upon thefe Infects, read before the Society of Naturalifts of Berlin, fays, That, in fome parts of the Eaft Indies, the queens are given alive to old men for ftrengthening the back, and that the natives have a method of catching the winged infects, which he calls females, before the time of emigration. They make two holes in the neft; the one to windward, and the other to leeward. At the leeward opening they place the mouth of a pot, previoufly rubbed within with an aromatic herb called *Bergera*, which is more valued there than the laurel in Europe. On the windward fide they make a fire of ftinking materials, which not only drives thefe infects into the pots, but frequently the hooded fnakes alfo, on which account they are obliged to be cautious in removing them. By this method they catch great quantities, of which they make with flour a variety of paftry, which they can afford to fell very cheap to the poorer ranks of people. Mr. KONIG adds, that in feafons when this kind of food is very plentiful, the too great ufe of it brings on an epidemic colic and dyfentery, which kills in two or three hours.

I have

On the following morning, however, as I have obſerved, they are to be ſeen running upon the ground in chace of each other

I have not found the Africans ſo ingenious in procuring or dreſſing them. They are content with a very ſmall part of thoſe which, at the time of ſwarming, or rather of emigration, fall into the neighbouring waters, which they ſkim off with calabaſhes, bring large kettles full of them to their habitations, and parch them in iron pots over a gentle fire, ſtirring them about as is uſually done in roaſting coffee. In that ſtate, without ſauce or any other addition, they ſerve them as delicious food ; and they put them by hands-full into their mouths, as we do com-fits. I have eat them dreſſed this way ſeveral times, and think them both delicate, nouriſhing, and wholeſome ; they are ſomething ſweeter, but not ſo fat and cloy-ing as the caterpillar or maggot of the *Palm-tree Snout-beetle*, *Curculio Palmarum*, which is ſerved up at all the luxurious tables of Weſt Indian epicures, particularly of the French, as the greateſt dainty of the Weſtern world.

According to the Baron DE GEER, Mr. SPARRMAN ſays, that the Hottentots eat theſe inſects, and even grow fat upon them; but does not ſay what methods they take to procure or dreſs them. DE GEER, *Memoires des Inſectes*, tom. VII. p. 49.

(22) PISO, DE LAET, MARCGRAVE, and other writers, mention their being an article of diet in different parts of South America.

" Alia præterea datur grandis ſpecies *Tama-ioura* dicta digiti articulum adæ-
" quans. Quarum etiam clunes deſſecantur et friguntur pro bono alimento."
PISO, Hiſt. Natural. lib. I. p. 9. lib. V. 291.

(23) MARCGR. Hiſt. Nat. 56.

(24) " Denique formicæ hic viſuntur grandiſſimæ, quas indigenæ vulgo come-
" dunt; et in foris venales habent." DE LAET. Americæ Utriuſque Deſcriptio,
p. 333.

" Formicis veſcebantur, eaſque ſtudioſe ad victum educabant. Ibid. p. 379."

(25) Sir HANS SLOANE ſays, the ſilk-cotton-tree worm is eſteemed by the Indians and negroes beyond marrow. This worm is no more than a large maggot, being the Caterpillar of a large Capricorn Beetle, or Goat Chafer : the Larva of a pretty large Cerambix (the *Lamia Tribulus* of FABRICIUS) which is alſo brought from Africa, where I have eaten thoſe worms roaſted. This inſect is moſt pro-bably to be found in all countries where the ſilk-cotton-tree (*Bombax*) is indi-genous. SLOANE's Jamaica, vol. II. p. 193.

I have

other; fometimes with one or two wings ftill hanging to their bodies, which are not only ufelefs, but feem rather cumber-fome [26].

The greater part have no wings, but they run exceeding faft, the males after the females; I have fometimes remarked two males after one female, contending with great eagernefs who fhould win the prize, regardlefs of the innumerable dangers that furrounded them.

They are now become from one of the moft active, induf-trious, and rapacious, from one of the moft fierce and impla-cable little animals in the world, the moft innocent, helplefs, and cowardly; never making the leaft refiftance to the fmalleft Ant. The Ants are to be feen on every fide in infinite numbers, of various fpecies and fizes, dragging thefe annual victims of the laws of nature to their different nefts. It is wonderful that a pair fhould ever efcape fo many dangers, and get into a place of fecurity. Some, however, are fo fortunate; and being found by fome of the labouring infects that are continually running about the furface of the ground under their covered galleries, which I fhall fhortly defcribe, are *elected* KINGS and QUEENS of new ftates; all thofe who are not fo elected and preferved certainly perifh, and moft probably in the courfe of the following day. The man-ner in which thefe labourers protect the happy pair from their innumerable enemies, not only on the day of the

I have difcourfed with feveral gentlemen upon the tafte of the white Ants; and on comparing notes we have always agreed, that they are moft delicious and deli-cate eating. One gentleman compared them to fugared marrow, another to fugared cream and a pafte of fweet almonds.

[26] LIGON obferved them, but does not know what they are. LIGON's Barba-does, p. 63.

maffacre of almoft all their race, but for a long time after, will I hope juftify me in the ufe of the term ELECTION. The little induftrious creatures immediately enclofe them in a fmall chamber of clay fuitable to their fize, into which at firft they leave but one fmall entrance, large enough for themfelves and the foldiers to go in and out, but much too little for either of the royal pair to make ufe of; and when neceffity obliges them to make more entrances, they are never larger; fo that, of courfe, the *voluntary fubjects* charge themfelves with the tafk of providing for the offspring of their fovereigns as well as to work and to fight for them until they fhall have raifed a progeny capable at leaft of dividing the tafk with them.

It is not until this time, probably, that they confummate their marriage, as I never faw a pair of them joined. The bufinefs of propagation, however, foon commences, and the labourers having conftructed a fmall wooden nurfery, as before defcribed, carry the eggs and lodge them there as faft as they can obtain them from the *queen.*

About this time a moft extraordinary change begins to take place in the *queen*, to which I know nothing fimilar, except in the PULEX PENETRANS of LINNÆUS, the JIGGER of the *Weft Indies*, and in the different fpecies of COCCUS, COCHINEAL. The abdomen of this female begins gradually to extend and enlarge to fuch an enormous fize, that an *old queen* will have it increafed fo as to be *fifteen hundred* or *two thoufand times* the bulk of the reft of her body, and *twenty or thirty thoufand times* the bulk of a labourer, as I have found by carefully weighing and computing the different ftates (tab. X. fig. 3.). The fkin between the fegments of the abdomen extends in every direction; and at laft the fegments are removed to half an

inch

inch diftance from each other, though at firft the length of the whole abdomen is not half an inch. They preferve their dark brown colour, and the upper part of the abdomen is marked with a regular feries of brown bars from the thorax to the pofterior part of the abdomen, while the intervals between them are covered with a thin, delicate, tranfparent fkin, and appear of a fine cream colour, a little fhaded by the dark colour of the inteftines and watery fluid feen here and there beneath. I conjecture the animal is upward of two years old when the abdomen is increafed to three inches in length : I have fometimes found them of near twice that fize. The abdomen is now of an irregular oblong fhape, being contracted by the mufcles of every fegment, and is become one vaft matrix full of eggs, which make long circumvolutions through an innumerable quantity of very minute veffels that circulate round the infide in a ferpentine manner, which would exercife the ingenuity of a fkilful anatomift to diffect and develope. This fingular matrix is not more remarkable for its amazing extenfion and fize than for its periftaltic motion, which refembles the undulating of waves, and continues inceffantly without any apparent effort of the animal ; fo that one part or other alternately is rifing and finking in perpetual fucceffion, and the matrix feems never at reft [27], but is always protruding eggs to the amount (as I have frequently counted in old queens) of fixty in a minute [28], or eighty thoufand and upward in one day of twenty-four hours [29]. Thefe

[27] " We may obferve in a *queen,* diftended with egg, a partition along the " back, and a continued motion from one extreme to the other, much like " that we find in filk-worms." Account of Englifh Ants by GOULD, p. 22.

[28] I cannot pofitively affert, that the old queens yield eggs fo plentifully at all

Z 2

Thefe eggs are inftantly taken from her body by her attendants (of whom there always are, in the royal chamber and
the galleries adjacent, a fufficient number in waiting) and carried to the nurferies, which in a great neft may fome of them
be four or five feet diftant in a ftraight line, and confequently
much farther by their winding galleries. Here, after they
are hatched, the young are attended and provided with
every thing neceffary until they are able to fhift for themfelves, and take their fhare of the labours of the community.
The foregoing, I flatter myfelf, is an accurate defcription
and account of the *Termes bellicofus* or fpecies that builds the
large nefts in its different ftates.

Thofe which build either the roofed turrets or the nefts in
the trees, feem in moft inftances to have a ftrong refemblance to
them, both in their form and oeconomy, going through the
fame changes from the egg to the winged ftate. The *queens*
alfo increafe to a great fize when compared with the labourers ;
but very fhort of thofe *queens* before defcribed. The largeft
are from about an inch to an inch and a half long, and not
much thicker than a common quill. There is the fame kind of
periftaltic motion in the abdomen, but in a much fmaller de-

times, but the protruding them being the confequence of the periftaltic motion,
it would feem involuntary on their parts, and the number, or nearly fo, always
indifpenfable : the aftonifhing multitudes of inhabitants found in their nefts alfo
countenance this opinion ftrongly.

(19) Since the reading of this paper, Mr. JOHN HUNTER, fo celebrated for his
great fkill and experience in comparative anatomy, has diffected two young queens.
He finds the abdomen contains two ovaria, in each of which are many hundred ova-
ducts, and in each of thefe ova-ducts a vaft many eggs ; fo that there feems no doubt
of the fact, as the matrix of a *full-grown queen* muft be calculated for the production
and yielding of a prodigious number of eggs. He has alfo diffected the kings ;
The refult of thefe diffections, with fome further particulars, will be related in
another paper.

gree; and, as the animal is incapable of moving from her place, the eggs no doubt are carried to the different cells by the labourers, and reared with a care fimilar to that which is prac-tifed in the larger nefts.

It is remarkable of all thefe different fpecies, that the work-ing and the fighting infects never expofe themfelves to the open air; but either travel under ground, or within fuch trees and fubftances as they deftroy, except, indeed, when they cannot proceed by their latent paffages, and find it convenient or ne-ceffary to fearch for plunder above ground. In that cafe they make pipes of that material with which they build their nefts. The larger fort ufe the red clay; the turret builders ufe the black clay; and thofe which build in the trees employ the fame ligneous fubftances of which their nefts are compofed (30) (31) (32).

With.

---

(30) " Small birds, fowls, Lizards, and other reptiles, fearch for them as the " moft delicious morfels; therefore they never go abroad but under their covered " ways.". DU TERTRE, quarto, vol. II. p. 345.

(31) " The earth hereabouts was all filled with a fpecies of a white Ant, called " Vag Vague, different from that which I have elfewhere defcribed. This, " inftead of raifing pyramids, continues buried under ground, and never makes " itfelf known but by fmall cylindrical galleries of the thicknefs of a goofe quill, " which it erects againft the feveral bodies it defigns to attack. Thefe galleries " are formed of earth with infinite delicacy of workmanfhip. The Vag Vagues " make ufe of them as of covert-ways, to work without being feen; and what-" ever they faften themfelves to, whether it be leather, cloth, linen, books, or " wood, it is furely gnawed and confumed. I fhould have thought myfelf pretty " well off, had they only attacked the reeds of my hut; but they pierced " through a trunk which ftood on treftles a foot above the ground, and gnawed " moft of my book." ADANSON's Voyage to Guinea, 179—337.

N. B. Mr. ADANSON is certainly miftaken when he fays, " They never make " themfelves known but by their covered ways, and is the only one whom I have
" met

With thefe materials they completely line moft of the roads leading from their nefts into the various parts of the country, and travel out and home with the utmoft fecurity in all kinds of weather. If they meet a rock or any other obftruction, they will make their way upon the furface ; and for that purpofe erect a covered way or arch, ftill of the fame materials, continuing it with many windings and ramifications through large groves ; having, where it is poffible, fubterranean pipes running parallel with them, into which they fink and fave themfelves, if their galleries above ground are deftroyed by any violence, or the tread of men or animals alarms them. When one chances by accident to enter any folitary grove, where the ground is pretty well covered with their arched galleries, they give the alarm by loud hiffings, which we hear diftinctly at every ftep we make; foon after which we may examine their galleries in vain for the infects, but find little holes, juft large enough for them, by which they have made their efcape into their fubterraneous roads. Thefe galleries are large enough for them to pafs and repafs fo as to prevent any ftoppages (though there are always numerous paffengers) and fhelter them equally from light and

---

" met with who has been attacked while living by the *white Ants*." I have fome doubt, that, although the approaches of the *Termites* were carried up to his bed, the bites he received were from *real Ants*, of which there are fome fcarce vifible which are very numerous and produce great pain ; whereas the bite of the *Termes* lets out much blood, and fhews not the leaft fymptom of venom. See DU TERTRE's Antilles, vol. II. p. 344. and Defcript. de l'Afrique, par LABAT, tom. III. p. 298.

(32) See SLOANE, LIGON, LINNÆUS (Termes Fatalis), FORSKAL (Termes Arda), and the various voyages to Africa and both Indies.

air, as well as from their enemies, of which the ants, being the moſt numerous, are the moſt formidable.

The *Termites,* except their heads, are exceeding ſoft, and covered with a very thin and delicate ſkin ; being blind, they are no match on open ground for the ants, who can ſee, and are all of them covered with a ſtrong horny ſhell not eaſily pierced, and are of diſpoſitions bold, active, and rapacious. Whenever the Termites are diſlodged from their covered ways, the various ſpecies of the former, who probably are as numerous above ground as the latter are in their ſubterraneous paſſages, inſtantly ſeize and drag them away to their neſts, to feed the young brood (33) (34) (35). The Termites are therefore exceeding ſolicitous

(33) Sir HANS SLOANE was certainly miſtaken in his account of the Wood Ants ; it is utterly improbable that they ſhould go into the neſts of the red Ants and kill them. It is moſt probable, the error has ariſen from Sir HANS's having confounded the two genera of inſects the *Formica* and *Termes* together, which made him never ſpeak of them with preciſion. The reverſe of his account is moſt likely, which is, that the *Formicæ* will follow their plunder into the neſts of the *Termites* and deſtroy them; for the latter always keep within their neſts or covered ways, avoiding all communications with other inſects and animals, and never meddling with them but when dead; whereas the *Formicæ* ramble about every where, and enter every cranny and hole that is large enough, and attack not only inſects and reptiles but even large animals. See SLOANE's Voyage to Jamaica, vol. II. p. 221, 222. tab. 238. *Hiſt. de l'Academie Royale des Sciences,* 1701, p. 16. *Fourmis de Viſite.*

(34) LIGON mentions another ſort of *Ants,* and deſcribes the galleries of the *Termites.* LIGON's Barbadoes, p. 64, 65.

(35) MERIAN ſays, the *Ants* make neſts above eight feet high, by which I apprehend ſhe means the neſts of the *Termites;* but in ſpeaking of the manners of the inſects ſhe certainly means ſome ſpecies of the *Formica.* Thoſe which are deſcribed as ſtripping the trees are a ſpecies called, in Tobago, *Para-ſol-Ants,* becauſe they cut out of the leaves of certain trees and plants pieces almoſt circular,

and

folicitous about the preferving their covered ways in good repair; and if you demolifh one of them, for a few inches in
length, it is wonderful how foon they rebuild it. At firft in
their hurry they get into the open part an inch or two, but ftop
fo fuddenly that it is very apparent they are furprized: for
though fome run ftraight on, and get under the arch as fpeedily
as poffible in the further part, moft of them run as faft back,
and very few will venture through that part of the track
which is left uncovered. In a few minutes you will perceive
them rebuilding the arch, and by the next morning they will
have reftored their gallery for three or four yards in length, if
fo much has been ruined; and upon opening it again will be
found as numerous as ever, under it, paffing both ways. If
you continue to deftroy it feveral times, they will at length feem
to give up the point, and build another in a different direction;
but, if the old one led to fome favourite plunder, in a few days
will rebuild it again; and, unlefs you deftroy their neft, never
totally abandon their gallery.

The *Termites arborum,* thofe which build in trees, frequently
eftablifh their nefts within the roofs and other parts of houfes,
to which they do confiderable damage, if not timely extir-
pated.

The large fpecies are, however, not only much more deftructive, but more difficult to be guarded againft, fince they
make their approaches chiefly under ground, defcending below
the foundations of houfes and ftores at feveral feet from the furface, and rifing again either in the floors, or entering at the

and are to be feen all the year round travelling from the plants along their road to
the neft, with each one of thefe circular pieces of leaves in their jaws, which,
from their fhape and colour, give a very good idea of people walking with para-
lofs (umbrellas). MERIAN, *Infectes de Surinam,* p. 18.

bottoms

bottoms of the pofts, of which the fides of the buildings are compofed, bore quite through them, followi..g the courfe of the fibres to the top, or making lateral perforations and cavities here and there as they proceed.

While fome are employed in gutting the pofts, others afcend from them, entering a rafter or fome other part of the roof. If they once find the thatch, which feems to be a favourite food, they foon bring up wet clay, and build their pipes or galleries through the roof in various directions, as long as it will fupport them ; fometimes eating the palm-tree leaves and branches of which it is compofed, and, perhaps (for variety feems very pleafing to them) the rattan or other running plant which is ufed as a cord to tye the various parts of the roof together, and that to the pofts which fupport it : thus, with the affiftance of the rats, who during the rainy feafon are apt to fhelter them-felves there, and to burrow through it, they very foon ruin the houfe by weakening the faftenings and expofing it to the wet. In the mean time the pofts will be perforated in every direction as full of holes as that timber in the bottoms of fhips which has been bored by the worms; the fibrous and knotty parts, which are the hardeft, being left to the laft [36].

They

<hr>

[36] The fea worms, fo pernicious to our fhipping, appear to have the fame office allotted them in the waters which the Termites have on the land. They will appear, on a very little confideration, to be moft important beings in the great chain of creation, and pleafing demonftrations of that infinitely wife and gracious Power which formed, and ftill preferves, the whole in fuch wonderful order and beauty : for if it was not for the rapacity of thefe and fuch animals, tropical rivers, and indeed the ocean itfelf, would be choked with the bodies of trees which are annually carried down by the rapid torrents, as many of them would laft for ages, and probably be productive of evils, of which, happily,

They fometimes, in carrying on this bufinefs, find, I will
not pretend to fay how, that the poft has fome weight to fupport, and then if it is a convenient track to the roof, or is itfelf
a kind of wood agreeable to them, they bring their mortar,
and fill all or moft of the cavities, leaving the neceffary roads
through it, and as faft as they take away the wood replace the
vacancy with that material; which being worked together by
them clofer and more compactly than human ftrength or art
could ram it, when the houfe is pulled to pieces, in order to
examine if any of the pofts are fit to be ufed again, thofe of the
fofter kinds are often found reduced almoft to a fhell, and all or
a greater part transformed from wood to clay as folid and as
hard as many kinds of free-ftone ufed for building in England..
It is much the fame when the *Termites bellicoſi* get into a cheft
or trunk containing cloaths and other things; if the weight

we cannot in the prefent harmonious ftate of things form any idea *; whereas now
being confumed by thefe animals, they are more eafily broken in pieces by the
waves; and the fragments which are not devoured become fpecifically lighter,
and are confequently more readily and more effectually thrown on fhore, where the
fun, wind, infects, and various other inftruments, fpeedily promote their entire
diffolution, and reftore the conftituent particles to that

―――――― " Mighty hand,
" Which, ever bufy, wheels the filent fpheres;
" Works in the fecret deep; fhoots, fteaming, thence
" The fair profufion that o'erfpreads the fpring:
" Flings from the fun direct the flaming day;
" Feeds every creature; hurls the tempeft forth;
" And, as on earth this grateful change revolves,
" With tranfport touches all the fprings of life."   THOMSON.

* That wood will endure in water an amazing number of ages, is apparent from the *oak ſtakes* which
we're driven into the bed of the river Thames on the invafion of this ifland by *Julius Cæfar*, one of which is
to be feen in Sir ASHTON LEVER's Mufeum, and likewife from thofe bodies of trees which are daily found
in the bogs and moraffes of Great Britain and Ireland; which after a duration, the former of eighteen
hundred, the latter of upwards of two thoufand years, are found in a perfect ftate of prefervation.

above is great, or they are afraid of Ants or other enemies, and have time, they carry their pipes through, and replace a great part with clay, running their galleries in various directions. The tree Termites, indeed, when they get within a box, often make a nest there, and being once in poſſeſſion deſtroy it at their leiſure. They did ſo to the pyramidal box which contained my compound microſcope. It was of mahogany, and I had left it in the ſtore of Governor CAMPBELL of Tobago, for a few months, while I made the tour of the Leeward Iſlands. On my return I found theſe inſects had done much miſchief in the ſtore, and, among other things, had taken poſſeſſion of the microſcope, and eaten every thing about it except the glaſs or metal, and the board on which the pedeſtal is fixed, with the drawers under it, and the things incloſed. The cells were built all round the pedeſtal and the tube, and attached to it on every ſide. All the glaſſes which were covered with the wooden ſubſtance of their neſts retained a cloud of a gummy nature upon them that was not eaſily got off, and the lacquer or burniſh with which the braſs work was covered was totally ſpoiled. Another party had taken a liking to the ſtaves of a Madeira caſk, and had let out almoſt a pipe of fine old wine. If the large ſpecies of Africa (the *Termites bellicoſi*) had been ſo long in the uninterrupted poſſeſſion of ſuch a ſtore, they would not have left twenty pounds weight of wood remaining of the whole building, and all that it contained [37].

Theſe

---

[37] Captain PHILLIP of the navy, who was ſome time at the Brazils in the ſervice of Portugal, gives me the following relation. " An engineer returned " from ſurveying the country, left his trunk on a table : the next morning, not " only all his cloaths were deſtroyed by *white Ants* or *Cutters*, but his papers alſo ; " and the latter in ſuch a manner, that there was not a bit left of an inch ſquare.

" The

Thefe infects are not lefs expeditious in deftroying the fhelves, wainfcotting, and other fixtures of an houfe, than the houfe itfelf. They are for ever piercing and boring in all directions, and fometimes go out of the broadfide of one poft into that of another joining to it ; but they prefer and always deftroy the fofter fubftances the firft, and are particularly fond of pine and fir-boards, which they excavate and carry away with wonderful difpatch and aftonifhing cunning : for, except a fhelf has fomething ftanding upon it, as a book, or any thing elfe which may tempt them, they will not perforate the furface, but artfully preferve it quite whole, and eat away all the infide, except a few fibres which barely keep the two fides connected together, fo that a piece of an inch-board which appears folid to the eye will not weigh much more than two fheets of pafteboard of equal dimenfions, after thefe animals have been a little while in poffeffion of it (38) (39) (40) (41). In fhort, the Termites are

fo

" The black lead pencils were likewife fo completely deftroyed, that the fmalleft
" piece, even of the *black lead* could not be found. The cloaths were not
" entirely cut to pieces and carried away, but appeared as if moth-eaten, there
" being fcarce a piece as large as a fhilling that was free from fmall holes ; and
" it was further remarkable, that fome *filver coin*, which was in the trunk, had a
" number of black fpecks on it, caufed by fomething fo corrofive that they could
" not eafily be rubbed off even with fand." Queen's-fquare, Wednefday, Jan.
17, 1781.

(38) " The white Ants are tranfparent as glafs, and bite fo forcibly, that in
" the fpace of one night alone they can eat their way through a thick wooden
" cheft of goods, and make it as full of holes, as if it had been fhot through
" with hail-fhot." BOSMAN's Guinea, p. 276, 7. 493.

(39) MOORE's Travels, p. 221.

(40) Voyage de LABAT aux Ifles, tom. II. p. 331.

(41) " The wood Ants are the moft pernicious of all others, being fo very
" deftructive to timber of moft forts, that, if not prevented, they will in a few
" years

fo infidious in their attacks, that we cannot be too much on our guard againft them : they will fometimes begin and raife their works, efpecially in new houfes, through the floor [42]. If you deftroy the work fo begun, and make a fire upon the fpot, the next night they will attempt to rife through another part ; and, if they happen to emerge under a cheft or trunk early in the night, will pierce the bottom, and deftroy or fpoil every thing in it before the morning [43]. On thefe accounts we are careful to fet all our chefts and boxes upon ftones or bricks, fo as to leave the bottoms of fuch furniture fome inches above the ground ; which not only prevents thefe infects finding them out fo readily, but preferves the bottoms from a corrofive damp which would ftrike from the earth through, and rot every thing therein : a vaft deal of vermin alfo would harbour under, fuch as Cock-roaches, Centipedes, Millepedes, Scorpions, Ants, and various other noifome infects.

When the Termites attack trees and branches in the open air, they fometimes vary their manner of doing it. If a ftake in a hedge has not taken root and vegetated, it becomes their bufi- nefs to deftroy it. If it has a good found bark round it, they

" years time deftroy the whole roof of an houfe, efpecially if it be of foft tim-
" ber. ——— They have likewife caufed great loffes to fhop-keepers, by boring
" through whole bales of linnen as well as woolen cloths. HUGHES's Barbadoes,
p. 93.

(42) The floors are generally made of the ftone or clay taken from the hills raifed by thefe infects, which, being moiftened with water, and mixed by treading, is beaten level, fmooth, and compact, with their feet and a kind of hand-bat or beetle.

(43) " One night, in a few hours, they pierced one foot of the table, and
" (having in that manner afcended) carried their arch acrofs it, and then
" down through the middle of the other foot into the floor, as good luck would
" have it, without doing any damage to the papers left there." KEMPFER Hift.
Japan, vol. II. p. 127.

will

will enter at the bottom, and eat all but the bark, which will remain, and exhibit the appearance of a folid ftick (which fome vagrant colony of Ants or other infects often fhelter in till the winds difperfe it) ; but if they cannot truft the bark, they cover the whole ftick with their mortar, and it then looks as if it had been dipped into thick mud that had been dried on. Under this covering they work, leaving no more of the ftick and bark than is barely fufficient to fupport it, and frequently not the fmalleft particle, fo that upon a very fmall tap with your walking ftick, the whole ftake, though apparently as thick as your arm, and five or fix feet long, lofes its form, and difappearing like a fhadow falls in fmall fragments at your feet. They generally enter the body of a large tree which has fallen through age or been thrown down by violence, on the fide next the ground, and eat away at their leifure within the bark, without giving themfelves the trouble either to cover it on the outfide, or to replace the wood which they have removed from within, being fomehow fenfible that there is no neceffity for it. Thefe excavated trees have deceived me two or three times in running : for, attempting to ftep two or three feet high, I might as well have attempted to ftep upon a cloud, and have come down with fuch unexpected violence, that, befides fhaking my teeth and bones almoft to diflocation, I have been precipitated, head foremoft, among the neighbouring trees and bufhes. Sometimes, though feldom, the animals are known to attack living trees; but not, I apprehend, before fymptoms of mortification have appeared at the roots, fince it is evident, as is before obferved, that thefe infects are intended in the order of nature to haften the diffolution of fuch trees and vegetables as have arrived at their greateft maturity and per-

5   fection,

fection, and which would, by a tedious decay, ferve only to encumber the face of the earth. This purpofe they anfwer fo effectually, that nothing perifhable efcapes them, and it is almoft impoffible to leave any thing penetrable upon the ground a long time in fafety; for the odds are, that, put it where you will abroad, they will find it out before the following morning, and its deftruction follows very foon of courfe. In confequence of this difpofition, the woods never remain long encumbered with the fallen trunks of trees or their branches; and thus it is, as I have before obferved, the total deftruction of deferted towns is fo effectually completed, that in two or three years a thick wood fills the fpace; and, unlefs *iron-wood* pofts have been made ufe of, not the leaft veftige of an houfe is to be difcovered.

The firft object of admiration which ftrikes one upon open-ing their hills is the behaviour of the foldiers. If you make a breach in a flight part of the building, and do it quickly with a ftrong hoe or pick-axe, in the fpace of a few feconds a foldier will run out, and walk about the breach, as if to fee whether the enemy is gone, or to examine what is the caufe of the attack. He will fometimes go in again, as if to give the alarm: but moft frequently, in a fhort time, is followed by two or three others, who run as faft as they can, ftraggling after one ano-ther, and are foon followed by a large body who rufh out as faft as the breach will permit them, and fo they proceed, the number increafing, as long as any one continues battering their building [44]. It is not eafy to defcribe the rage and

fury

(44) " They throw up little hills of feven or eight feet high, fo very full of
" holes that they rather feem like honey-combs than burrows. Thefe Ant hills
" are of a very fmall circumference in proportion to their height, being fharp at top,
" fo that to judge by the looks of them one would think the wind could blow them
" down.

fury they fhew. In their hurry they frequently mifs their
hold, and tumble down the fides of the hill, but recover themfelves as quickly as poffible; and, being blind, bite every
thing they run againft, and thus make a crackling noife, while
fome of them beat repeatedly with their forceps upon the building, and make a fmall vibrating noife, fomething fhriller and
quicker than the ticking of a watch : I could diftinguifh this
noife at three or four feet diftance, and it continued for a minute
at a time, with fhort intervals. While the attack proceeds they
are in the moft violent buftle and agitation. If they get hold of
any one, they will in an inftant let out blood enough to weigh
againft their whole body ; and if it is the leg they wound, you
will fee the ftain upon the ftocking extend an inch in width.
They make their hooked jaws meet at the firft ftroke, and never
quit their hold, but fuffer themfelves to be pulled away leg by
leg, and piece after piece, without the leaft attempt to efcape.
On the other hand, keep out of their way, and give them no
interruption, and they will in lefs than half an hour retire into
the neft, as if they fuppofed the wonderful monfter that da-
maged their caftle to be gone beyond their reach. Before they are
all got in you will fee the labourers in motion, and haftening
in various directions toward the breach : every one with
a burthen of mortar in his mouth ready tempered. This they
ftick upon the breach as faft as they come up, and do it with
fo much difpatch and facility, that although there are thoufands, and I may fay millions, of them, they never ftop or

" down. I one day attempted to knock off the top of one of them with my cane,
" but the ftroke had no other effect than to bring fome thoufands of the animals
" out of doors, to fee what was the matter: upon which I took to my heels and
" ran away as faft as I could." SMITH's Voyage to Guinea.

embarrafs

embarrafs one another; and you are moft agreeably deceived
when, after an apparent fcene of hurry and confufion, a regular wall arifes, gradually filling up the chafm. While they are
thus employed, almoft all the foldiers are retired quite out of
fight, except here and there one, who faunters about among fix
hundred or a thoufand of the labourers, but never touches the
mortar either to lift or carry it; one, in particular, places himfelf clofe to the wall they are building. This foldier will turn
himfelf leifurely on all fides, and every now and then, at intervals of a minute or two, lift up his head, and with his forceps
beat upon the building, and make the vibrating noife before
mentioned; on which immediately a loud hifs, which appears
to come from all the labourers, iffues from within fide the
dome and all the fubterraneous caverns and paffages: that it
does come from the labourers is very evident, for you will fee
them all haften at every fuch fignal, redouble their pace, and
work as faft again.

As the moft interefting experiments become dull by repetition
or continuance, fo the uniformity with which this bufinefs is
carried on, though fo very wonderful, at laft fatiates the mind.
A renewal of the attack, however, inftantly changes the fcene,
and gratifies our curiofity ftill more. At every ftroke we hear a
loud hifs; and on the firft the labourers run into the many pipes
and galleries with which the building is perforated, which they
do fo quickly that they feem to vanifh, for in a few feconds all
are gone, and the foldiers rufh out as numerous and as vindictive
as before (45). On finding no enemy they return again leifurely
into

(45) By the foldiers being fo ready to run out upon the repetition of the
attack, it appears, that they but juft withdraw out of fight, to leave room for
the labourers to proceed without interruption in repairing the breach, and in this

into the hill, and very foon after the labourers appear loaded as
at firft, as active and as fedulous, with foldiers here and there
among them, who act juft in the fame manner, one or other
of them giving the fignal to haften the bufinefs. Thus the
pleafure of feeing them come out to fight or to work alternately may be obtained as often as curiofity excites or time
permits : and it will certainly be found, that the one order never
attempts to fight, or the other to work, let the emergency be
ever fo great.

We meet vaft obftacles in examining the interior parts of
thefe tumuli. In the firft place, the works, for inftance,
the apartments which furround the royal chamber and the
nurferies, and indeed the whole internal fabric, are moift,
and confequently the clay is very brittle : they have alfo fo
clofe a connexion, that they can only be feen as it were by
piece-meal ; for having a kind of geometrical dependance or
abutment againft each other, the breaking of one arch pulls down
two or three. To thefe obftacles muft be added the obfti-
nacy of the foldiers, who fight to the very laft, difputing
every inch of ground fo well as often to drive away the
negroes who are without fhoes, and make white people
bleed plentifully through their ftockings. Neither can we
let a building ftand fo as to get a view of the interior parts
without interruption, for while the foldiers are defending the

---

inftance they fhew more good fenfe than the bulk of mankind, for, in cafe of a con-
flagration in a city, the number of people who affemble to ftare is much greater
than of thofe who come to affift, and the former always interrupt and hinder the
latter in their efforts. The fudden retreat of the labourers, in cafe of an alarm,
is alfo a wonderful inftance of good order and difcipline, feldom feen in populous
cities, where we frequently find helplefs people, women, and children, without
any ill intention, intermixing in violent tumults and dangerous riots.

                                         out-works,

out-works, the labourers keep barricadoing all the way againſt us, ſtopping up the different galleries and paſſages which lead to the various apartments, particularly the royal chamber, all the entrances to which they fill up ſo artfully as not to let it be diſtinguiſhable while it remains moiſt; and externally it has no other appearance than that of a ſhapeleſs lump of clay [46]. It is, however, eaſily found from its ſituation with reſpect to the other parts of the building, and by the crouds of labourers and ſoldiers which furround it, who ſhew their loyalty and fidelity by dying under its walls. The royal chamber in a large neſt is capacious enough to hold many hundreds of the attendants, beſides the royal pair, and you always find it as full of them as it can hold. Theſe faithful ſubjects never abandon their charge even in the laſt diſtreſs; for whenever I took out the royal chamber, and, as I often did, preſerved it for ſome time in a large glaſs bowl, all the attendants continued running in one direction round the king and queen with the utmoſt ſolicitude, ſome of them ſtopping on every circuit at the head of the latter, as if to give her ſomething. When they came to the extremity of the abdomen, they took the eggs from her, and carried them away, and piled them carefully together in ſome part of the chamber, or in the bowl under, or behind any pieces of broken clay which lay moſt convenient for the purpoſe.

[46] In tab. VIII. fig. 2. and 4. the entrances of the royal chamber, now exhibited, are repreſented open. They were all ſhut by the labourers before I had got to it, and were opened ſince I arrived in England. Two or three of them, however, are not quite open in the chamber itſelf, and that next the breach at A, and marked with a croſs ⊕, is ſtill left ſhut, as a ſpecimen of the manner in which they do it. I have alſo more royal chambers and various ſpecimens of the interior buildings, with ſeveral galleries and paſſages, ſhut up while we were attacking the neſt.

Some

Some of thefe little unhappy creatures would ramble from
the chamber, as if to explore the caufe of fuch a horrid ruin
and cataftrophe to their immenfe building, as it muft appear to
them ; and, after fruitlefs endeavours to get over the fide
of the bowl, return and mix with the croud that continue
running round their common parents to the laft (tab. VIII. fig.
4. B.). Others, placing themfelves along her fide, get hold of
the queen's vaft matrix with their jaws, and pull with all their
ftrength fo as vifibly to lift up the part which they fix at ; but,
as I never faw any effect from thefe attempts, I never could
determine whether this pulling was with an intention to remove
her body, or to ftimulate her to move herfelf, or for any other
purpofe ; but, after many ineffectual tugs, they would defift
and join in the croud running round, or affift fome of thofe
who are cutting off clay from the external parts of the chamber or fome of the fragments and moiftening it with the
juices of their bodies, to begin to work a thin arched fhell
over the body of the queen, as if to exclude the air, or to hide
her from the obfervation of fome enemy. Thefe, if not in-
terrupted, before the next morning, completely cover her,
leaving room enough within for great numbers to run about her.

I do not mention the king in this cafe, becaufe he is very fmall
in proportion to the queen, not being bigger than thirty of the
labourers, fo that he generally conceals himfelf under one fide
of the abdomen, except when he goes up to the queen's head,
which he does now and then, but not fo frequently as the reft.

If in your attack on the hill you ftop fhort of the royal chamber, and cut down about half of the building, and leave open
fome thoufands of galleries and chambers, they will all be fhut
up with thin fheets of clay before the next morning. If even
the whole is pulled down, and the different buildings are thrown

4   in

in a confused heap of ruins, provided the king and queen are not deftroyed or taken away, every interftice between the ruins, at which either cold or wet can poffibly enter, will be fo covered as to exclude both, and, if the animals are left undifturbed, in about a year they will raife the building to near its priftine fize and grandeur.

The marching Termites are not lefs curious in their order, as far as I have had an opportunity of obferving them, than thofe defcribed before. This fpecies feems much fcarcer and larger than the *Termes bellicofus.* I could get no information relative to them from the black people, from which I conjecture they are little known to them : my feeing them was very accidental. One day, having made an excurfion with my gun up the river Camerankoes, on my return through the thick foreft, whilft I was fauntering very filently in hopes of finding fome fport, on a fudden I heard a loud hifs, which, on account of the many ferpents in thofe countries, is a moft alarming found. The next ftep caufed a repetition of the noife, which I foon recognized, and was rather furprifed feeing no covered ways or hills. The noife, however, led me a few paces from the path, where, to my great aftonifhment and pleafure, I faw an army of Termites coming out of a hole in the ground, which could not be above four or five inches wide. They came out in vaft numbers, moving forward as faft feemingly as it was poffible for them to march. In lefs than a yard from this place they divided into two ftreams or columns, compofed chiefly of the firft order, which I call labourers, twelve or fifteen a-breaft, and crouded as clofe after one another as fheep in a drove, going ftraight forward without deviating to the right or left. Among thefe, here and there, one of the foldiers was to be feen, trudging along with them, in the fame manner, neither ftopping

or

or turning; and as he carried his enormous large head with apparent difficulty, he put me in mind of a very large ox amidft a flock of fheep.   While thefe were buftling along, a great many foldiers were to be feen fpread about on both fides of the two lines of march, fome a foot or two diftant, ftanding ftill or fauntering about as if upon the look out leaft fome enemy fhould fuddenly come upon the labourers.   But the moft extraordinary part of this march was the conduct of fome others of the foldiers, who having mounted the plants which grow thinly here and there in the thick fhade, had placed themfelves upon the points of the leaves, which were elevated ten or fifteen inches above the ground, and hung over the army marching below. Every now and then one or other of them beat with his forceps upon the leaf, and made the fame fort of ticking noife which I had fo frequently obferved to be made by the foldier who acts the part of a furveyor or fuper-intendant when the labourers are at work repairing a breach made in one of the common hills of the *Termites bellicofi*.   This fignal among the marching white Ants produced a fimilar effect; for, whenever it was made, the whole army returned a hifs, and obeyed the fignal by increafing their pace with the utmoft hurry.   The foldiers who had mounted aloft, and gave thefe fignals, fat quite ftill during the intervals (except making now and then a flight turn of the head) and feemed as folicitous to keep their pofts as regular centinels. The two columns of the army joined into one about twelve or fifteen paces from their feparation, having in no part been above three yards afunder, and then defcended into the earth by two or three holes. They continued marching by me for above an hour that I ftood admiring them, and feemed neither to increafe or diminifh their numbers, the foldiers only excepted, who

quitted

quitted the line of march,   nfelves at different
diftances on each fide of the two columns; for they appeared
much more numerous before I quitted the fpot. Not expect-
ing to fee any change in their march, and being pinched for
time, the tide being nearly up, and our departure fixed at
high water, I quitted the fcene with fome regret, as the obfervation of a day or two might have afforded me the opportunity
of exploring the reafon and neceffity of their marching with
fuch expedition, as well as of difcovering their chief fettlement,
which is probably built in the fame manner as the large hills
before defcribed. If fo, it may be larger and more curious, as
thefe infects were at leaft one-third larger than the other fpecies,
and confequently their buildings muft be more wonderful if
poffible: thus much is certain, there muft be fome fixed place
for their king and queen, and the young ones. Of thefe fpecies I have not feen the perfect infect.

The œconomy of nature is wonderfully difplayed in a comparative obfervation on the different fpecies who are calculated
to live under ground until they have wings, and this fpecies
which marches in great bodies in open day. The former, in
the two firft ftates, that is, of labourers and foldiers, have no
eyes that I could ever difcover; but when they arrive at the
winged or perfect ftate in which they are to appear abroad,
though only for a few hours, and that chiefly in the night,
they are furnifhed with two confpicuous and fine eyes:
fo the *Termes viarum*, or marching Bugga Bugs, being intended to walk in the open air and light, are even in the firft
ftate furnifhed with eyes proportionably as fine as thofe
which are given to the winged or perfect infects of the
other fpecies.

I am

I am afraid of e~~n~~ ~~ac~~   ~~a~~ your time, which leads me
to drop the fubject for the prefent; but, as my materials are
not exhaufted, if thefe fheets meet with your approbation, it
will encourage me to give fome further particulars, with ob-
fervations and reflections, at a future period.

I have the honour to be, &c.

Henry Smeathman del.

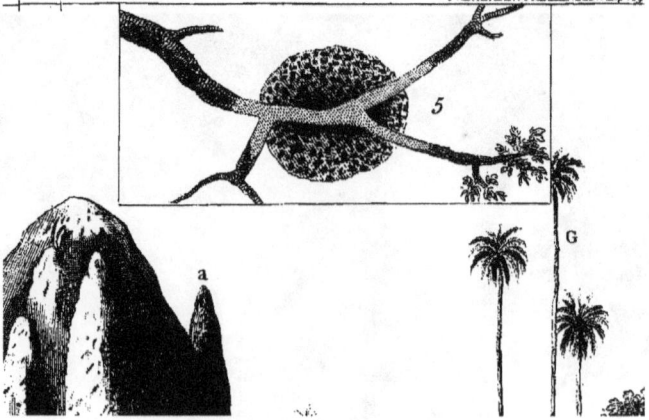

b.VII.p.192.

3

Basire Sc.

1

2

3

4

b

b

a   *a*

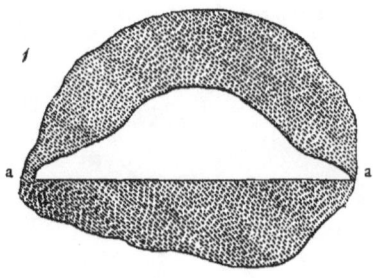

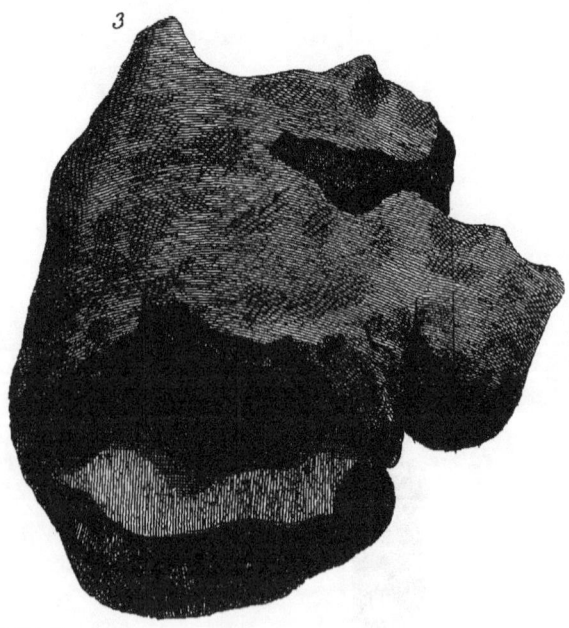

Henry Smeathman del.

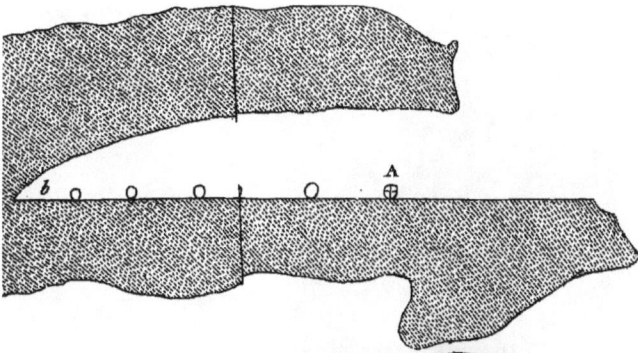

5.

7

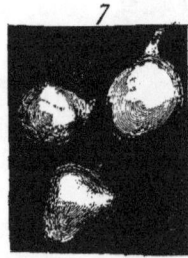

6

Basire Sc.

Henry Smithman Del.

5

Basire Sc

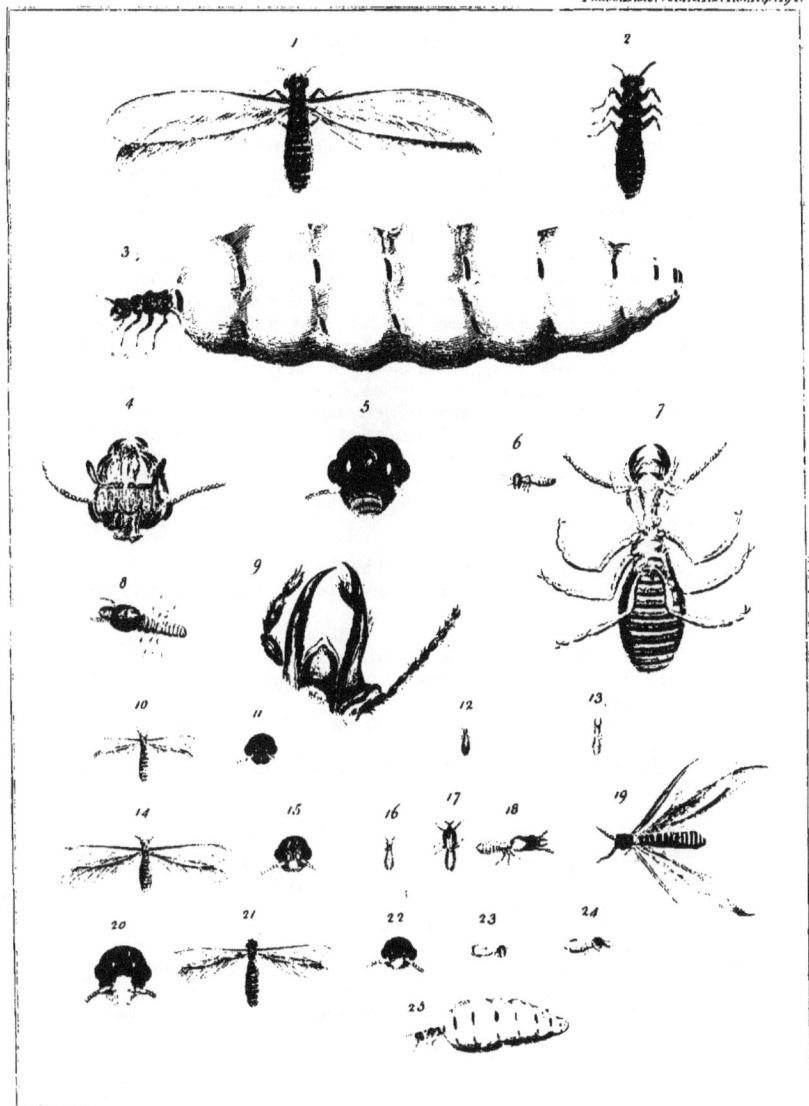

Explanation of the plates to Mr. SMEATHMAN's Account of the Termites of Africa, &c.

TAB. VII. fig. 1. The hill-neft raifed by the Termites bellicofi, defcribed page 148.
  aaa. Turrets by which their hills are raifed and enlarged, p. 150.
Fig. 2. A feƈtion of fig. 1. as it would appear on being cut down through the middle from the top a foot lower than the furface of the ground, p. 154.
AA. An horizontal line from A on the left, and a pe pendicular line from A at the bottom, will in erfeƈt each other at the royal chamber, p. 154.
  The darker fhades near it are the empty apartments and paffages, which it feems are left fo for the attendants on the king and queen, who, when old, may require near one hundred thoufand to wait on them every day.
  The parts which are the leaft fhaded and dotted are the nurferies, furroun !ed, like the royal chamber by empty paffages on all fides for the more eafy accefs to them with the eggs from the queen, the provifion for the young, &c. N. B. The magazines of provifions are fituated without any feeming order among the vacant paffages which furround the nurferies.
  B. The top of the interior building, which often feems, from the arches carrying upward, to be adorned on the fides with pinnacle , p. 156.
  C. The floor of the area or nave, p. 156.
  DDD. The large galleries which afcend from under all the buildings fpirally to the top, p. 156.
  EE. the bridges, p. 158.
Fig. 3. The firft appearance of an hill-neft by two turrets, p. 150.
Fig. 4. A tree, with the neft of the Termites arborum, and their covered way, p. 161.
  FFFF. Covered ways of the Termites arborum, p. 173.
Fig. 5. A feƈtion of the neft of the Termites arborum.
Fig. 6. A neft of the Termites bellicofi, with Europeans on it, feemingly obferving a veffel at fea, p. 151.
Fig. 7. A bull ftanding centinel upon one of thefe nefls, while the reft of the herd is ruminating below, p. 151.
  GGG. The African palm-trees, from the nuts of which is made the Oleum Palmæ.

Tab. VIII. fig. 1. A tranfverfe feƈtion of a royal chamber, p. 151.
  aa. The t in fides in which the entrances are made, p. 152.
Fig. 2. A longitudinal feƈtion of a royal chamber, p. 151.
  b. The entrances, p. 187.
  A. The door fhut up, as left by the labourers, p. 187.
Fig. 3. A royal chamber fore-fhortened.
Fig. 4. the fame royal chamber reprefented as juft opened, and difcovering (b) the queen, and her attendants running round her, p. 188.
  bb. A line drawn from *b* to *b* will run along the range of doors or entrances, p. 187.
  AAA. A line run from A to AA will crofs the door, which remains clofed as it was found. The reft are reprefented as they appear fince the mortar,
with

with which they were flopped up, has been in part or wholly picked out with a small instrument, p. 187.

Fig. 5. A nursery, p. 153.

Fig. 6. A little nursery, with the eggs, the young ones, the mushrooms, mouldinefs, &c. as just taken from the hill, p. 153.

Fig. 7. The mushrooms magnified by a strong lens, p. 154.

Tab. IX. fig. 1. and 2. The turret nefts, with roofs of the Termes mordax and a Term s atrox as finished, p. 159.

Fig. 3. A turret, with the roof beg n.

Fig. 4. A turret, raised only about half i's height.

Fig. 5. A turret, building upon one which had been thrown down, p. 160.

Fig. 6. 6. A turret, broken in two.

Tab. X. fig. 1. A Termes bellicofus, p. 141. numb. 1. and p. 165.

Fig. 2. A KING. N.B. A king never alters his form after he lofes his wings, neither does he apparently increafe in bulk.

Fig. 3. A QUEEN, p. 170.

Fig. 4. The head of a perfect infect magnified.

Fig. 5. A face, with flemmata magnified, p. 141. numb. 1.

Fig. 6. A labourer, p. 163.

Fig. 7. A labourer magnified.

Fig. 8. A foldier, p. 164.

Fig. 9. A foldi r's forceps and part of his head magnified, p. 164.

Fig. 10. The Termes mordax, p. 141. numb. 2. and p. 161.

Fig. 11. The face with the flemmata magnified, p. 141. numb. 2.

Fig. 12. A labourer.

Fig. 13. A foldier.

Fig. 14. The Termes atrox, p. 141. numb. 3. and p. 160.

Fig. 15. The face and flemmata magnified, p. 141. numb. 3.

Fig. 16. A labourer.

Fig. 17. A foldier.

Fig. 18. Idem.

Fig. 19. The Termes deftructor, p. 141. numb. 4.

Fig. 20. The face and flemmata magnified, p. 141. numb. 4.

Fig. 21. The Termes arborum, p. 141. numb. 5. and p. 162.

Fig. 22. The face and flemmata magnified, p. 141. numb. 5.

Fig. 23. A labourer.

Fig. 24. A foldier.

Fig. 25. A QUEEN, p. 172.

N. B. In the figures 5. 11. 15. 20. and 21. the two white fpots between the edges are the flemmata.

www.ingramcontent.com/pod-product-compliance
Lightning Source LLC
Chambersburg PA
CBHW022005190326
41519CB00010B/1389